kjbnf KO
560.1792 LOH-H

Loh-Hagan, Virginia, author
The ice age
33410015689260   02-06-2021

Kouts Public Library
101 E. Daumer Road
Kouts, IN 46347

Surviving History

THE ICE AGE

Virginia Loh-Hagan

45th Parallel Press

Published in the United States of America by Cherry Lake Publishing Group
Ann Arbor, Michigan
www.cherrylakepublishing.com

Reading Adviser: Marla Conn, MS, Ed., Literacy specialist, Read-Ability, Inc.
Book Designer: Melinda Millward

Photo Credits: © Esteban De Armas/Shutterstock.com, front cover, 1; © DanielMcHenry/Shutterstock.com, 4; © Marko Aliaksandr/Shutterstock.com, 6; © Procy/Shutterstock.com, 8; © Gorodenkoff/Shutterstock.com, 10; © MariskaVegter/Shutterstock.com, 12; © sekarb/iStock.com, 14; © Vitalii Bashkatov/Shutterstock.com, 16; © Esteban De Armas/Shutterstock.com, 18, back cover; © deepak bishnoi/Shutterstock.com, 20; © EcoPrint/Shutterstock.com, 22; © Daniel Eskridge/Shutterstock.com, 24, 27; © Christopher Wood/Shutterstock.com, 28

Graphic Element Credits: © Milos Djapovic/Shutterstock.com, back cover, front cover; © cajoer/Shutterstock.com, back cover, front cover, multiple interior pages; © GUSAK OLENA/Shutterstock.com, back cover, multiple interior pages; © Miloje/Shutterstock.com, front cover; © Rtstudio/Shutterstock.com, multiple interior pages; © Konstantin Nikiteev/Dreamstime.com, 29

Copyright © 2021 by Cherry Lake Publishing Group
All rights reserved. No part of this book may be reproduced or utilized in any form or by any means without written permission from the publisher.
45TH Parallel Press is an imprint of Cherry Lake Publishing Group.

Library of Congress Cataloging-in-Publication Data

Names: Loh-Hagan, Virginia, author.
Title: The ice age / Virginia Loh-Hagan.
Description: Ann Arbor, Michigan : Cherry Lake Publishing, [2021] | Series: Surviving history | Includes index.
Identifiers: LCCN 2020030324 (print) | LCCN 2020030325 (ebook) | ISBN 9781534180307 (hardcover) | ISBN 9781534182011 (paperback) | ISBN 9781534181311 (pdf) | ISBN 9781534183025 (ebook)
Subjects: LCSH: Climatic changes—Juvenile literature.
Classification: LCC QC981.8.C5 L64 2021 (print) | LCC QC981.8.C5 (ebook) | DDC 560/.1792–dc23
LC record available at https://lccn.loc.gov/2020030324
LC ebook record available at https://lccn.loc.gov/2020030325

Cherry Lake Publishing Group would like to acknowledge the work of the Partnership for 21st Century Learning, a Network of Battelle for Kids. Please visit http://www.battelleforkids.org/networks/p21 for more information.

Printed in the United States of America
Corporate Graphics

TABLE OF CONTENTS

INTRODUCTION 4
MOVE OR DIE? 8
WARM OR COLD? 12
GATHER OR HUNT? 16
SHELTER OR STORMS? 20
FLIGHT OR FIGHT? 24
SURVIVAL RESULTS 28
DIGGING DEEPER: DID YOU KNOW...? 30

Glossary 32
Learn More! 32
Index 32
About the Author 32

INTRODUCTION

The La Brea Tar Pits have preserved animal bones for thousands of years. Preserve means to keep as it was originally.

An ice age is a period of millions of years. It's when Earth is really cold. Earth is covered with large ice sheets. It's covered with mountain **glaciers**. Glaciers are large bodies of moving ice. They thaw and refreeze over time. Within an ice age, there are many periods of warmer weather. Earth is currently in a period of warmth during an ice age. The Antarctic and Greenland ice sheets are still here.

The last major ice age was during the Pleistocene **Epoch**. Epochs are periods of time. This epoch began about 2.6 million years ago. It lasted until about 11,700 years ago. One of the best sources of information about the Pleistocene Epoch is the La Brea Tar Pits in Los Angeles, California.

Some experts think a **comet** may have wiped out the large animals. A comet is a space object made of ice and dust.

During this epoch, the continents were where they are now. The weather was colder and drier. Most of the water was ice. There was little rain.

Early humans appeared at the end of the epoch. They spread around the world. They thrived. They became the dominant land animal. They hunted large animals.

These large animals had a hard time during periods of freezing weather. Many of these animals became **extinct**. Extinct means no longer found alive.

MOVE OR DIE?

Some experts think today's humans came from the group that **evolved** in South Africa. Evolve means to change over time.

Sea levels were lower. There were land bridges. Early humans and animals could **migrate**. Migrate means to move from one area to another.

The weather was freezing cold. Plants couldn't grow. Animals needed plants to eat. They moved to warmer areas where plants still grew. Early humans needed to eat animals. They followed the animals.

Some experts think early humans survived by living in South Africa. They called this area the "Garden of **Eden**." Eden means paradise. These experts found human bones and **artifacts** in caves. Artifacts are things made by humans. This area had a lot of food resources.

QUESTION 1

Would you have migrated or not?

A You embraced the **nomadic** life. Nomads are people who move around. You had no home base. You moved to find food. You moved often. You were always on the go.

B You moved north and south. When it got too cold, you moved south. When it got too warm, you moved back north. You had two home bases.

C You didn't want to move. You wanted to stay in one place. You didn't want to leave your home. You were scared of changes.

Modern humans spread out. They replaced all other human species.

SURVIVOR BIOGRAPHY

Most top **predators** died out in the last ice age. Predators hunt other animals for food. For example, dire wolves became extinct. But gray wolves survived. They were smaller than dire wolves. This meant they could move faster and easier. Gray wolves could hunt small animals. This meant they had a lot of food sources. They hunted and lived in packs. They talked to each other by howling and barking. This meant they could defend themselves against larger predators. They were widely spread. They lived in different habitats. They traveled over large areas to hunt. They **adapted**. Adapt means to change. They changed as needed. As such, they've been around for the last 300,000 years. They have roamed North America, Europe, and Asia. Today, they're the largest living carnivores in North America. Carnivores are meat eaters. Gray wolves are mostly found in Canada and the Northwestern United States. Modern wolves and dogs can trace their history to the late Pleistocene gray wolves.

WARM OR COLD?

Cold climates provided challenges for early humans.

Early humans had to survive cold **climates**. Climate means weather. If not, they'd die. They could get **frostbite**. Frostbite happens when body parts freeze. They could fall off. Early humans could get **hypothermia**. Hypothermia happens when bodies lose heat.

Early humans adapted to harsh climates. Early humans made tools. Tools helped them survive.

Early humans made fire. Fire provides heat. It kept early humans warm.

They used animal furs. They used animal bones. They made bone needles. They made stone scrapers to scrape the furs. They used the needles to sew furs together. They made clothes. They used clothes to stay warm.

QUESTION 2

How would you have kept warm?

A You made coats with fur lining. You made coats with fur **trimming**. Trimming is the edges on clothes. It protected against cold winds. You added trimming to the hood. You added trimming to the sleeves. You mostly used wolverine fur. Your coat was a mix of long and short hairs. It was a tight fit. This kept you really warm.

B You made capes. You could wrap the cape around you. But you were still exposed to the cold weather. Your capes were made from cows and deer.

C You wore animal skins of large animals. You put these skins around your shoulders.

Some early humans made necklaces from eagle talons. Talons are bird claws.

SURVIVAL TIPS

The Antarctic ice sheet is the largest mass of ice on Earth. It's an accumulation of Antarctic snowfall over millions of years. Follow these tips to survive in Antarctica:

- Keep your entire body covered. Cover the tip of your nose. Wind can freeze your skin on contact.
- Wear at least 4 layers of clothes. Keep your clothes dry.
- Drink lots of water. The air is dry. You'll lose water by breathing.
- Eat fatty foods. You need a lot of calories to stay warm.
- Always carry a survival bag. Pack a tent, water, food, and a stove. Also pack an ice shovel. Pack a communication device.
- Do exercises to warm up. Swing your arms to keep blood flowing to your hands.
- Watch for "umbles." When people mumble or stumble, they may be freezing to death.
- Prepare for a whiteout. A whiteout means you can't see anything.

GATHER OR HUNT?

Sometimes, early humans had to walk a long time to find food.

Early humans lived in small groups. They collected food for each other. They lived off the land. If there was plenty of food, they stayed. They protected the land. If food ran out, they migrated to different lands.

They gathered plants. They ate grasses. They ate fruits. They ate seeds. They ate nuts.

They hunted animals. They did this instead of eating meat left behind by predators. Predators hunt **prey**. Prey are animals used for food.

They made hunting tools. They sharpened stones for cutting. They made wooden spears. They made stone tips. They made bows and arrows. They made bone fishhooks.

QUESTION 3

How would you have gotten food?

A You were a gatherer. Gatherers collected plants to eat. You knew which plants to eat. You knew which plants not to eat. Some plants could make you sick.

B You lived by water. You were a fisher. You scooped fish with your hands. You waited for fish to wash ashore. You used bone fishhooks.

C You were a hunter. You watched animal predators hunt. You learned from them. You used hunting tools to track prey. You killed animals. You brought them back to the group.

Early humans taught their children hunting and gathering skills.

SURVIVAL BY THE NUMBERS

- There have been at least 5 major ice ages during the 4.6 billion years of Earth's history.
- During the Pleistocene Epoch, there were at least 20 warm and cold phases.
- During the Pleistocene Epoch, about 32 percent of Earth's land was covered with glacial ice. Right now, about 10 percent of Earth's land is covered with glacial ice. Glacial ice includes glaciers, ice caps, and ice sheets.
- The last of the cave bears became extinct around 25,000 to 28,000 years ago.
- Cave lions and woolly rhinoceroses became extinct around 14,000 years ago.
- Woolly mammoths became extinct around 3,600 years ago.
- Antarctica has been partially covered by an ice sheet for the past 40 million years. It has had 38 ice ages in the past 5 million years. Its winds blow over 100 miles (161 kilometers) per hour.

SHELTER OR STORMS?

Some of the first shelters were caves.

The Ice Age climate was harsh. There were extreme changes in weather. There were many storms. There were ice storms.

Early humans had to seek **shelter**. Shelter is a place that offers protection. Early humans needed protection from weather. They also needed protection from predators. They needed a place to sleep. Early nomadic humans relied on nature for shelter. They lived in caves.

Early humans started to settle down. They made more **permanent** homes. Permanent means lasting a long time. Some humans made their own shelters. They made huts or tents. They made them using wood, leaves, rocks, bones, and animal skins.

QUESTION 4

What type of living space would you have had in the Ice Age?

A You lived in caves. You were protected from the weather. You made fires to stay warm. But you had to be careful of predators. Cave bears and cave lions lived in caves.

B You lived under overhanging cliffs. You'd be protected from winds and rains. But you would still be exposed.

C You made your own shelter. It took a long time. You had to figure out a way to transport building materials. You had to be willing to stay in one place for a while. Your group members might leave without you.

Some early humans made cave paintings.

SURVIVAL TOOLS

Fire is important for human survival. It sets humans apart from animals. Fire needs fuel, oxygen, and a heat source. There must be plant life to have fuel. The air must have oxygen. Before humans, the main heat source was lightning strikes. About 1.5 million years ago in Africa, humans worked with fire. Proof was found in the Wonderwerk Cave. Evidence of regular use of fire was found in caves in Israel. This was 300,000 to 400,000 years ago. Early humans used fire for light. It was used for warmth. It was used to frighten off predators. Its smoke kept away bugs. Fire was also used for cooking. Learning to cook made humans smarter. Cooked meat gave humans more energy. Early humans made hearths. Hearths are fireplaces. Humans gathered around hearths. They learned to talk to each other. About 7,000 years ago, humans used fire to clear land. They used fire in wars.

FLIGHT OR FIGHT?

Saber-toothed tigers had a loud roar and are thought to have hunted in packs.

Megafauna lived during the Ice Age. Megafauna means giant animals. The Pleistocene megafauna were big for a reason. Their large sizes allowed them to stay warm in cold climates.

Woolly mammoths were the most famous. They were related to elephants. They were 14 feet (4.3 meters) tall. They weighed 12,000 pounds (5,443 kilograms). They had thick coats of brown hair. Their hair could be 3 feet (1 m) long. They had fur in their ears. They had long **tusks**. Tusks are long, pointed teeth.

Saber-toothed tigers did sneak attacks. Their teeth were 11 inches (28 centimeters) long.

QUESTION 5

What would you have done if you saw Pleistocene megafauna?

A You would flee. You had no fighting or hunting skills. You had no desire to mess with these giant animals. You stayed out of their way.

B You were a hunter. You had your hunting tools. One large animal could provide enough food for a whole group. You had to work with a group of hunters. The animals were too big to take down alone.

C You had to fight for protection. You may not have had tools with you.

Big animals were in the land, air, and water.

SURVIVAL RESULTS

Glaciers store about 69 percent of the world's freshwater.

Would you have survived?

Find out! Add up your answers to the chapter questions. Did you have more **A**s, **B**s, or **C**s?

- If you had more **A**s, then you're a survivor! Congrats!

- If you had more **B**s, then you're on the edge. With some luck, you might have just made it.

- If you had more **C**s, then you wouldn't have survived.

Are you happy with your results? Did you have a tie? Sometimes fate is already decided for us. Follow the link below to our webpage. Scroll until you find the series name *Surviving History*. Click download. Print out the template. Follow the directions to create your own paper die. Read the book again. Roll the die to find your new answers. Did your fate change?

https://cherrylakepublishing.com/teaching_guides

DIGGING DEEPER: DID YOU KNOW...?

The Ice Age was exciting and harsh. Surviving history involves many different factors. Dig deeper. Consider some of the facts below.

QUESTION 1:

Would you have migrated or not?
- The Bering Land Bridge connected Asia to the Americas.
- The Bering Land Bridge also connected Siberia and Alaska. That's how many animals came to North America.
- Humans evolved. They developed bigger brains. They developed flexible hands.

QUESTION 2:

How would you have kept warm?
- Early humans were cold during winter. They had evolved to lose most of their body hair.
- For early humans, clothes were used to stay warm. They weren't fashion.
- Clothes let early humans save heat energy. This let them look for food.

QUESTION 3:

How would you have gotten food?
- Over time, groups interacted with each other. This created trade.
- Adding more meat to their diets made early humans smarter.
- As humans evolved, their teeth got smaller.

QUESTION 4:

What type of living space would you have had in the Ice Age?
- Early humans lived near cave entrances. There was more daylight there. It was also warmer than deeper parts of the cave.
- Early humans camped along the shores of lakes and rivers.
- Early humans built huts in forests.

QUESTION 5:

What would you have done if you saw Pleistocene mcgafauna?
- There were also giant bears, giant sloths, and giant beavers. Giant beavers were the size of today's black bears.
- Oceans widened. Sharks and whales thrived.
- Some animals that survived the Ice Age include crocodiles, deer, and cockroaches.

GLOSSARY

adapted (uh-DAPT-id) changed as needed
artifacts (AHR-tuh-fakts) things made by humans
climates (KLYE-mits) weather conditions
comet (KAH-mit) a space object made of ice and dust with a tail
Eden (EE-den) a place of paradise
epoch (EP-uhk) period of time
evolved (ih-VAHLVD) changed over time
extinct (ik-STINGKT) no longer found alive
frostbite (FRAWST-bite) the freezing of the skin
glaciers (GLAY-shurz) large bodies of slow-moving ice
hypothermia (hye-puh-THUR-mee-uh) happens when bodies lose heat, which can lead to death

megafauna (MEG-uh-faw-nuh) big animals
migrate (MYE-grate) to move from one area to another
nomadic (no-MAD-ik) moving around from place to place
permanent (PUR-muh-nuhnt) lasting a long time
predators (PRED-uh-turz) animals that hunt other animals for food
prey (PRAY) animals that are hunted for food
shelter (SHEL-tur) a place that offers protection from weather or other dangers
trimming (TRIM-ing) edges on clothes
tusks (TUHSKS) long, pointed teeth

LEARN MORE!

- Hoena, Blake, and Alessandro Valdrighi (illus.). *Could You Survive the Ice Age? An Interactive Prehistoric Adventure.* North Mankato, MN: Capstone Press, 2020.
- Loh-Hagan, Virginia. *Origins of Life.* Ann Arbor, MI: Cherry Lake Publishing, 2020.
- Medina, Nico. *What Was the Ice Age?* New York, NY: Penguin Workshop, 2017.
- Tite, Jack. *Mega Meltdown: The Weird and Wonderful Animals of the Ice Age.* New York, NY: Blueprint Editions, 2018.

INDEX

adaptation, 11, 13
animals, 11, 19, 22, 24, 25, 26, 27, 31
Antarctic ice sheet, 15, 19

caves, 20, 21, 22, 23, 31
climate, 12, 13, 21
comets, 6

evolution, 8, 30, 31
extinction, 7

fire, 13, 21, 22, 23
fishing, 18
food, 16, 17, 18, 31

gatherers, 17, 18
glaciers, 5, 19, 28

humans, 7, 8, 9, 10, 13, 17
hunting, 17, 18, 26

Ice Age, 19
    animals (See animals)

digging deeper, 30–31
introduction, 4–7
survivor biography, 11

La Brea Tar Pits, 4, 5

megafauna, 25, 26, 27, 31
migrate, 9, 10, 17, 30

plants, 9, 17, 18, 23
Pleistocene Epoch, 5, 7, 11, 19

predators/prey, 11, 17, 21, 22

shelters, 20, 21, 22, 31
South Africa, 8, 9
survival tips, 15
survival tools, 23

wolves, 11

ABOUT THE AUTHOR

Dr. Virginia Loh-Hagan is an author, university professor, and former classroom teacher. She loves ice in her drinks. She lives in San Diego with her very tall husband and very naughty dogs. To learn more about her, visit www.virginialoh.com.